T0224591

essentials

Springer essentials

Springer essentials provide up-to-date knowledge in a concentrated form. They aim to deliver the essence of what counts as "state-of-the-art" in the current academic discussion or in practice. With their quick, uncomplicated and comprehensible information, essentials provide:

- an introduction to a current issue within your field of expertis
- an introduction to a new topic of interest
- an insight, in order to be able to join in the discussion on a particular topic

Available in electronic and printed format, the books present expert knowledge from Springer specialist authors in a compact form. They are particularly suitable for use as eBooks on tablet PCs, eBook readers and smartphones. *Springer essentials* form modules of knowledge from the areas economics, social sciences and humanities, technology and natural sciences, as well as from medicine, psychology and health professions, written by renowned Springer-authors across many disciplines.

More information about this subseries at http://www.springer.com/series/16761

Cordula Harter

Gluten Sensitivity

About Gluten-Associated Disorders
and the Purpose of a Gluten-Free
Diet

 Springer

Cordula Harter
Heidelberg University Biochemistry Center
Heidelberg, Germany

ISSN 2197-6708   ISSN 2197-6716 (electronic)
essentials
ISSN 2731-3107   ISSN 2731-3115 (electronic)
Springer essentials
ISBN 978-3-658-32656-2   ISBN 978-3-658-32657-9   (eBook)
https://doi.org/10.1007/978-3-658-32657-9

Responsible Editor: Sarah Koch
This Springer imprint is published by the registered company Springer Fachmedien Wiesbaden GmbH part of Springer Nature.
The registered company address is: Abraham-Lincoln-Str. 46, 65189 Wiesbaden, Germany

What You Can Find in This *essential*

- The main sources of gluten and what hides behind the term "gluten"
- What other ingredients of wheat can cause sensitivity
- The role of intestinal health and intestinal microbiota in wheat sensitivity
- Which disorders wheat and other gluten-containing grains can cause
- Arguments for and against a gluten-free diet

Preface

The number of people who complain of gluten sensitivity and consider wheat to be toxic or at least not beneficial to health has been rising steadily for years. The number of patients with a diagnosed gluten sensitivity is also rising. However, there is a discrepancy between the perceived and the medically proven gluten sensitivity.

Is gluten even the disease-causing agent? If so, why? What other substances in wheat can cause disorders? Does wheat deserve its "bad reputation"?

These questions and the observation that there is a large gap between knowing and believing, and that misbelief is spreading despite many facts, prompted me to take a scientifically -based approach to the subject.

"Gluten sensitivity" is almost a buzzword that sums up an extremely complex and heterogeneous issue. Complex and heterogeneous is not only the composition of gluten but also the intolerance, which can only in rare cases be clearly attributed to gluten. In many cases, the term "wheat sensitivity" would be more appropriate, but "gluten sensitivity" seems more in keeping with the spirit of the times.

There are clear diagnostic criteria for the medical evidence of gluten sensitivity. However, gluten sensitivity is often diagnosed by the people affected themselves without meeting known medical diagnostic criteria. It is undoubtedly important to take symptomatic patients seriously. New diagnostic criteria may still have to be found for many cases. However, it is clear that gluten sensitivity is following a trend that is nourished by the belief in disease-causing foods and from which the food industry earns a lot of money.

In this *essential* I bridge the gap between gluten-containing cereals and the mechanisms of disease development. I present the currently most discussed molecules and mechanisms that can cause gluten sensitivity—or better wheat sensitivity—and describe the known disease patterns.

The information in this *essential is* mainly based on articles in internationally renowned scientific journals, which I searched predominantly via the literature database "Pubmed."

I have tried to describe the facts in such a way that any scientifically interested person can understand them. For experts and those who would like to delve deeper into the subject, a list of current, scientific literature is provided.

I hope that every reader will find answers to his or her questions about gluten sensitivity in this *essential* and would be pleased if I could contribute to a factually sound discussion of this scientifically, medically, and socially important topic.

Cordula Harter

Contents

Introduction

1

Diarrhea, loss of appetite, and weight loss led to the diagnosis of "celiac disease" more than 100 years ago (Losowsky 2008). Initially, celiac disease was described as a disease of the abdomen and maldigestion due to its clinical symptoms. It was not until the 1950s that gluten, a mixture of storage proteins from various types of grain, was identified as the cause of celiac condition.

Today, gluten is associated with various disorders - which may not only affect the gastrointestinal tract but the entire organism (Felber et al. 2014).

"Gluten sensitivity" is the common term used to describe symptoms that occur after ingesting certain cereals, primarily wheat. From a medical point of view, a distinction is made between the following disorders associated with gluten: Celiac disease, wheat allergies, non-celiac non-wheat allergy-wheat sensitivity (Dale et al. 2019). These disorders can be distinguished from each other on the basis of diagnostic criteria or, in the absence of diagnostic criteria, on the basis of symptoms.

The triggers of gluten sensitivity and the underlying mechanisms of disease development are only partially known. Only in some clinical presentations gluten has been identified as the disease-causing agent. But gluten is not a uniform substance. Rather, gluten consists of hundreds of different proteins and is found in varying composition and quantity in wheat, rye, and barley.

Most of the proteins contained in gluten belong to the superfamily of prolamins, the most allergen-rich protein family (Juhasz et al. 2018). However, non-gluten proteins, especially the amylase trypsin inhibitors (ATIs) found in wheat, can also contribute to the development of disease (Schuppan and Zevallos 2015).

In addition to proteins, indigestible carbohydrates, so-called fermentable oligo-, di-, and monosaccharides and polyols (FODMAP) have been associated with reduced well-being after the consumption of cereals. Especially in cases of

© Springer Fachmedien Wiesbaden GmbH, part of Springer Nature 2021   1
C. Harter, *Gluten Sensitivity,* Springer essentials,
https://doi.org/10.1007/978-3-658-32657-9_1

medically unproven gluten sensitivity, FODMAP may account for the symptoms (Priyanka et al. 2018).

All disorders caused by wheat ingredients have in common that they affect the proper function of the gastrointestinal tract, resulting in digestive disorders and acute or chronic inflammation. The symptoms may spread from the gastrointestinal tract to other organ systems, resulting in far-reaching implications not only for the physical but also for mental health.

In some cases, eliminating wheat or gluten from the diet can help. In other cases, a more comprehensive change in diet can help. In any case, if gluten sensitivity is suspected, medical advice should be sought before making any long-term changes in dietary habits.

Cereals containing gluten are an important food in our culture: They are an important source of energy, which nourishes both humans and their intestinal bacteria, and have outstanding qualities in baking and cooking. In addition, gluten-containing cereals and cereal products contain valuable fiber, vitamins, and minerals. Without a medical need, the benefits of a gluten-free diet are questionable and there is a risk of malnutrition.

Gluten-free nutrition is a fashionable trend: Much more people consume gluten-free foods than there are patients medically clearly diagnosed with gluten or wheat sensitivity. The demand for gluten-free products nourishes a lucrative market and in many cases benefits the companies involved more than the consumer.

This *essential* explains on a molecular level which components of cereals and especially of wheat are possible triggers of disease. It explains the role of the intestine and the intestinal microbiota in the development of the disease. Celiac disease, wheat allergy, and non-celiac non-wheat allergy wheat sensitivity are distinguished from each other and possible mechanisms (or hypotheses) of the respective disease development are described.

Finally, nutritional aspects of a gluten-containing and gluten-free nutrition are discussed and the trend of the gluten-free market is analyzed.

Gluten and Wheat

2

2.1   Gluten Sources

Gluten is found in sweet grasses (Poaceae) of the subfamily Pooideae. These are divided into Triticeae, which include wheat (genus *Triticum*), barley (*Hordeum*), and rye (*Secale*), and Avenae, to which oats belong. Depending on the definition and country, oats are considered gluten-free or gluten-containing (Codex Alimentarius). However, the gluten-like proteins contained in oats, the avenins, are generally well tolerated by gluten-sensitive individuals. The gluten content in commercial oats often comes from contamination with wheat, barley, or rye.

Of the Triticeae, wheat contains the highest gluten content. However, the gluten content depends not only on the plant genus and variety but also on the environmental conditions in which the plant grows and the postharvest processing (Shewry et al. 2013). For this reason, data of the gluten content are often averages of several different varieties from different growing areas or from different grain processing operations (Table 2.1).

Gluten-free cereals are corn, rice, millet, and the "pseudo-cereals"—buckwheat, quinoa, and amaranth.

People with gluten sensitivity should avoid eating foods containing gluten. In order not to pose a risk to their health, they depend on reliable information about the absence of gluten. The definition of gluten-free foods is laid down in international regulations (Codex Alimentarius; Commission Implementing Regulation (EU) No 828/2014).

Definition of Gluten-Free Foods According to Codex Alimentarius 118-1979
Foods are considered **gluten free** if they do not contain wheat (applies to all types of wheat, such as durum wheat, khorasan wheat, emmer, einkorn, spelt), barley,

© Springer Fachmedien Wiesbaden GmbH, part of Springer Nature 2021
C. Harter, *Gluten Sensitivity,* Springer essentials,
https://doi.org/10.1007/978-3-658-32657-9_2

Table 2.1 Gluten content of some cereals and cereal products

Grain/grain product	Gluten content (mg/100 g food)
Spelt[a]	8100–11500[*]
Spelt flour type 630[b]	10,300
Spelt flour whole grain[b]	9500
Barley[b,c]	4200–5600[*]
Barley beads[b]	4700
Oats[b,c]	1300–4600[*]
Durum wheat flour whole grain, old varieties (before 1960)[d]	11,800
Durum wheat flour whole grain, new varieties (2004–2014)[d]	8500
Rye[b,c]	3100
Rye flour wholemeal[b]	3400
Wheat[a]	4900–13,700[*]
Wheat flour type 630[b]	9400
Wheat flour wholemeal[b]	8300
Wheat beer[b]	274
Light full beer[b]	3

[*]The variation is due to the fact that different varieties have been analyzed or that the analyses have been carried out by different laboratories
[a](Schalk et al. 2017a);
[b](Andersen et al. 2015);
[c](Schalk et al. 2017b);
[d](Ficco et al. 2019)

rye or oats* or their crossbred varieties, and if their **gluten content** (of the product sold to the consumer) **does not exceed 20 mg/kg.** Or, which consist of one or more ingredients from wheat (applies to all types of wheat, such as durum wheat, khorasan wheat, emmer, einkorn, spelt), barley, rye or oats* or their crossbred varieties, which have been specially processed to remove gluten, and the gluten content (of the product sold to the consumer) does not exceed 20 mg/kg

*Oats can be tolerated by most people who are intolerant to gluten. To what extent oats are permitted is subject to national regulations. According to EU regulation 828/2014, "Oats contained in a food presented as "gluten free" must have been specially produced, prepared, and/or processed in such a way to avoid contamination

by wheat, barley, rye, or their crossbred varieties and the gluten content of such oats cannot exceed 20 mg/kg."

Gluten-free foods that are not naturally gluten free, such as baked goods or pasta, are often marked with the symbol of a crossed grain. The crossed grain ensures that international regulations are met and that the food contains less than 20 mg gluten/kg. The gluten-free symbol is particularly useful for food preparations in which gluten is not suspected, such as fruit preparations thickened with wheat starch. Foods containing gluten in which gluten is not suspected must be declared.

2.2 Wheat

The wheat genus includes the wheat species einkorn, emmer, durum wheat, common or bread wheat, and spelt (Table 2.2). They are all genetically closely related (Bickel 2015).

Einkorn is genetically the simplest type of wheat. It has a diploid genome distributed over seven chromosomes ($2n = 14$, genome A). Emmer and durum wheat are tetraploid and contain twice as many chromosomes as Einkorn. The genome consists of two related sub-genomes ($2n = 28$, genome AB).

Table 2.2 Names and genomes of different wheat species

	Diploid series	Tetraploid series	Hexaploid series
Wild forms	Wild spelt (*Triticum boeoticum* or *T. monococcum* subsp. *aegilopoides*)	Wild emmer (*Triticum turgidum* subsp. *dicoccoides*)	?
Cultivated forms – Hulled wheat – Unhulled wheat	Einkorn (*T. monococcum*)	Emmer (*T. turgidum* subsp. *dicoccon*) Khorasan wheat, Kammut® (*T. turgidum* subsp. *turanicum*) Durum wheat (*T. turgidum* subsp. *durum*)	Spelt (*T. aestivum* subsp. *spelta*) Common wheat or bread wheat (*T. aestivum*)
Chromosomes	$2n = 14$	$2n = 4x = 28$	$2n = 6x = 42$
Genome	AA	AABB	AABBDD

Einkorn and emmer belong to the hulled cereals, that is, the grain is surrounded by a firm hull, the spelt. Since the hull does not fall off during threshing, it must be removed in an additional operation before the grain is processed. Emmer is also sold under the registered trade name Kamut®. It is a subspecies of emmer, the khorasan wheat, which was originally cultivated in modern-day Iran. Kamut® is a protected variety which must not be modified and must be grown according to certain ecological guidelines. Khorasan wheat requires dry and warm climate, so that it is hardly cultivated in Central Europe. Among the tetraploid wheat varieties, durum wheat, which is the result of modern breeding, plays the most important role economically. It is free-threshing and is mainly used for the production of pasta.

Common wheat and spelt are genetically the most complex types of wheat. They are hexaploid, that is, they contain 3 related subgenomes (2n = 42, genome ABD), which are distributed over 2 times 7 chromosomes each. The world's most important type of wheat is unhulled common or bread wheat, which is particularly high-yielding and of which there are about 200 different varieties available for cultivation in Germany alone (Federal Association of German Plant Growers eV—(Bundesverband deutscher Pflanzenzüchter e. V.)).

In the course of evolution, genomes change so that the plant can adapt to the prevailing environmental conditions. A change in a genome is a natural process of selection. In addition, plant genomes have been and are modified by targeted breeding, for example, to make the plant more stable or more resistant to climatic fluctuations or pathogens, and also to improve baking properties and yields. In breeding, different varieties are crossed with each other in order to obtain new varieties that have the best characteristics of both original varieties. Several thousand different wheat varieties, which have been created by conventional breeding methods, are approved worldwide. In contrast, no genetically modified wheat is currently approved for commercial cultivation and marketing in Europe or North America. However, there are field trials with genetically modified wheat on strictly designated areas in numerous countries. Genetically modified wheat developed in Germany is currently being cultivated in field trials in Switzerland. No field trials are carried out in Germany (Forum Bio- and Gene Technology—(Forum Bio- und Gentechnologie e. V.)).

How did the Genome of Bread Wheat Evolve?

The genome of bread or common wheat is one of the most complex genomes. It consists of the 3 subgenomes A, B, and D. How did this "megagenome" evolve, which consists of 15×10^9 base pairs and is about five times larger than the human genome? Of the wheat species that have been used as crops for thousands of years, bread wheat and spelt are the youngest.

About 7 million years ago, genomes A and B diverged from a single precursor genome. They contain the genetic information of the diploid original plants AA (einkorn) and BB (wild grass). Over time, the two A and B genomes hybridized to give rise to the tetraploid AABB genome (emmer). It is assumed that the D genome evolved about 5.5 million years ago from the A and B genomes. It still characterizes the plant *Aegilops tauschii,* the goat grass. Less than 400,000 years ago, *Aegilops tauschii* (DD genome) and emmer (AABB genome) hybridized to form the hexaploid wheat, the "primordial mother" of bread wheat (Marcussen et al. 2014).

The domestication of wheat, that is, the modification of wild plants to improve their quality and yield and to adapt them to environmental conditions, began about 10,000 years ago with the transition of the human lifestyle from hunter-gatherers to farmers and cattle breeders. It had its origin in the so-called fertile crescent, in which today's countries Egypt, Iraq, Iran, Jordan, Lebanon, Palestine, Saudi Arabia, and Turkey are located, and continued over the following 5000 years in a northwesterly direction to Europe.

Spelt—The Better Wheat?

Spelt often has a better reputation than wheat. Some consumers consider it to be better tolerated and many believe it is an ancient, original wheat variety. Genetically speaking, spelt belongs to the bread wheat and not to the species einkorn and emmer, which are sometimes referred to as "original wheat." The origin of spelt is controversially discussed. However, there are genetic data which suggest that spelt is a young wheat species which evolved by natural crossbreeding of domesticated emmer with bread wheat (Dvorak et al. 2012). Spelt was first cultivated in Europe about 6000 years ago (Lobitz 2018).

Because of some unfavourable traits for commercial processing and, accordingly, a small area under cultivation, spelt has undergone less manipulations by breeding methods than bread wheat in recent decades. Among the agronomic disadvantages of spelt are the plant's lack of stability, lower yield than bread wheat, and the hull that has to be removed before processing the grain. The baking properties of spelt are also highly dependent on the variety and require special care in dough preparation. However, in recent years the demand for spelt products and the cultivation of spelt has increased considerably.

Spelt is a relatively unpretentious and disease-resistant plant that thrives in the Central European climate and is ideal for organic cultivation. To compensate for its disadvantages, some spelt varieties have been crossed with bread wheat. However, new spelt varieties have also been created through varietal breeding. In spelt products the variety does not have to be indicated. Thus, the consumer does not know whether he is consuming pure spelt or a cross between spelt and bread wheat. Whether it is a pure or crossed spelt variety can be determined in the laboratory by biochemical analysis of certain gluten proteins (Wieser). Averaged over several varieties, spelt contains more gluten than normal bread wheat (Andersen et al. 2015). There is no scientific explanation for a better tolerance of spelt.

2.3   Gluten

There are different definitions of gluten, which vary according to purpose and target group. The European Union's definition is short and concise: "Gluten is a protein fraction of wheat, rye, barley, oats or their crossbreeds and derivatives which some people cannot tolerate and which is insoluble in water and 0.5 M sodium chloride solution" (EU implementing regulation No. 828/2014). The definitions are more comprehensive in a scientific context. Biologically, gluten fulfils the function of a storage protein in cereal grains and is found in wheat in the highest concentrations compared to other cereals (Schalk et al. 2017b).On a molecular basis, gluten proteins are encoded in all three subgenomes (ABD) of all wheat species (International Wheat Genome Sequencing Consortium 2018). Biochemically, gluten is a mixture of proteins from the prolamine family and the glutelin family (Wieser 2007). By treating gluten with aqueous ethanol, the prolamines can be dissolved, leaving behind the alcohol-insoluble glutelins. Glutelins form networks that can only be dissolved by adding reducing agents (which cleave the disulfide bridges within and between the individual glutelin molecules). Individual gluten proteins consist of 300–800 building blocks, the amino acids.

Alcohol-soluble prolamins of wheat are called gliadins, glutelins of wheat are called glutenins. Gluten in barley is formed by various hordeins. Gluten in rye is formed by secalins and gluten in oats of avenins (Schalk et al. 2017b) (Fig. 2.1).

Gluten is also known as a glue-like protein. It allows the formation of a cohesive, viscoelastic dough from flour by adding water, which can enclose many gas molecules, so that a voluminous bread can be produced. During this process, the glutelins form a stable scaffold of high-molecular protein complexes, which is responsible for the elasticity of the dough. The gliadins are deposited in the spaces between the scaffold, making the dough viscous, that is, soft and smooth (Scherf and Koehler 2016) (Fig. 2.2).

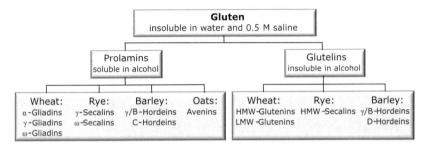

Fig. 2.1   Composition of gluten from wheat, rye, barley, and oats. HMW high molecular weight, LMW low molecular weight

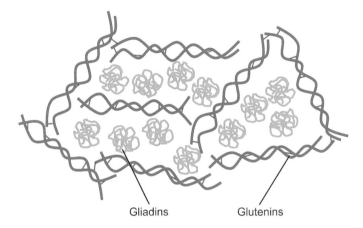

Gliadins   Glutenins

Fig. 2.2   Gluten scaffold of glutenins (blue) and gliadins (green). Glutenins consist of several protein chains, which are connected by disulfide bridges (red)

Gluten from different cereal genera, but also from different species within a genus, differs in the protein composition and structure of the different proteins. This different quality of gluten explains not only the differences in the baking properties of flours, but also the differences in tolerance and the effects on human health. In terms of baking properties, flours made from bread wheat produce easily workable, particularly stretchable doughs with a good gas retention capacity and thus large-volume breads. Flours from other cereals sometimes bind less water or can trap fewer gas molecules, making the breads more compact and heavier.

In quantitative terms, gluten accounts for about 75–80 % of the protein content of wheat grains (Schalk et al. 2017b). If a person suffers from wheat sensitivity, in most cases he or she also suffers from gluten sensitivity. However, gluten sensitivity can be attributed to a few of the gluten proteins. Therefore, it is ultimately the quality rather than the quantity of gluten supplied with the diet that causes the sensitivity (Schalk et al. 2017a).

Wheat

Wheat contains the highest gluten content of all cereals. However, the quality of gluten, that is, the composition of different gluten proteins, is more important for tolerance than quantity. Wheat varieties that contain the D genome—such as bread wheat and spelt—contain more potentially harmful gluten proteins than wheat varieties that only contain the A and B genomes, such as einkorn and emmer. However, old varieties of einkorn and emmer, which have not undergone major breeding modification, also contain gluten.

Gluten Peptides

Gluten peptides are formed by the enzymatic cleavage of gluten proteins into smaller fragments. At present, a database (Allergen Online) lists more than 1000 peptides derived from gluten proteins of wheat, barley, rye, and oats that can be potentially harmful to individuals with celiac disease. These peptides have been identified by the sequence of certain amino acids and their ability to induce an immune response. However, the fact that they have immunogenic properties does not mean that they actually cause symptoms in celiac patients or individuals with gluten sensitivity. Many of the peptides have very similar properties and there is an overlap of amino acid sequences.

Some gluten peptides are described as toxic, others as immunogenic. Both terms are often used synonymously, but "toxic" generally describes a damaging effect of a molecule without necessarily causing an immune reaction. "Immunogenic" means that a molecule is capable of eliciting a specific immune response, which often depends on the genetic disposition of the individual. Immune reactions can be measured by determining, for example, inflammatory mediators or antibodies in the blood. In contrast, toxic reactions often lack a specific, measurable parameter.

Gluten peptides differ from peptides of other food proteins mainly by the following characteristics:

Digestibility

Gluten peptides are formed in the gastrointestinal tract through the cleavage of gluten proteins by digestive enzymes. Since there are hundreds of different gluten proteins and each protein consists of several hundred amino acids, tens of thousands of different gluten peptides can be produced. Normally, food peptides are largely broken down into individual amino acids during digestion, which are absorbed by the intestinal epithelial cells and released into the blood. However, gluten peptides behave differently: They resist the cleavage by digestive enzymes.

Potential Triggers of Inflammatory Reactions
Gluten peptides are harmless in people who are not sensitive to gluten: They can simply be excreted with the stool. They can also be utilized by intestinal bacteria (Herran et al. 2017). In people with celiac disease or gluten sensitivity, gluten peptides can cause damage: They can pass through the intestinal epithelium and trigger an inflammatory reaction in the underlying tissue layer (Chap. 4).

Special Structure
Gluten peptides are characterized by a high content of the amino acids proline (P) and glutamine (Q). However, the human digestive system is not equipped with enzymes that can break down proline-rich peptides. Therefore, the only way to get rid of them is to excrete them or to have intestinal bacteria help to utilize them.

   The unique composition with many prolines and glutamines is also used for the specific detection of gluten peptides, for example in the food industry, in clinical chemistry or in research laboratories. For this purpose, certified tests (AgraQuant® Gluten G12 and RIDASCREEN®) are available on the market, which allow the quantitative determination of the gluten content in a sample using antibodies. AgraQuant® Gluten G12 recognizes the peptide QPQLPY (L, leucine; Y, tyrosine), RIDASCREEN® recognizes QQPFP (F, phenylalanine).

   Among the potentially most toxic and immunocompetent gluten proteins are the α-gliadins of wheat. It is estimated that bread wheat contains between 25 and 150 copies of various α-gliadins (Ozuna et al. 2015). Since the genes for α-gliadins occur on all 3 wheat genomes, the proteins are also present in einkorn and emmer. However, the largest number of immunogenic α-gliadins is encoded in the D genome.

   Two peptides of the α-gliadins are regarded as particularly toxic or immunogenic: p31–43, which contains the amino acids 31–43, is believed to be responsible for damaging the barrier function of the intestinal epithelium (Fasano 2011), and the so-called 33mer, which consists of a specific sequence of 33 amino acids (Fig. 2.3). It resists digestion in the intestinal lumen and reaches the connective tissue layer of the intestine, where it triggers a specific immune reaction in patients with celiac disease, a so-called T-cell response. Since only short peptides of 7–9 amino acids are required for the activation of T-cells, the 33mer contains several recognition sequences, so-called T-cell epitopes (Fig. 2.3).

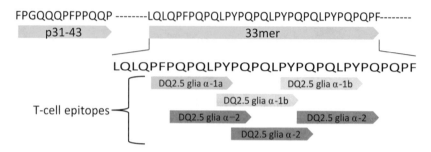

Fig. 2.3   Partial amino acid sequence of α-gliadin. p31–43 damages the barrier function of the small intestine. 33mer triggers a specific immune response in patients with celiac disease via its T-cell epitopes. F, phenylalanine; G, glycine; L, leucine; P, proline; Q, glutamine; Y, tyrosine

33mer

33mer, the most immunogenic gluten peptide, is present in the largest amounts in α-gliadins encoded on the D genome. It is, therefore, found in larger quantities in bread wheat and spelt than in einkorn, emmer, or other cereals. However, the amount of 33mer in hexaploid wheat also depends on the variety. In one study 40 wheat varieties were investigated, including 2 spelt varieties. The content of 33mer varied from 0.03–0.6 mg/g of flour. In rye flour the content of 33mer was below the detection limit (Schalk et al. 2017a).

Although gluten peptides resist human digestion, bacterial and plant enzymes can break them down into individual amino acids. This means that gluten proteins can be completely broken down if they get in contact with these enzymes under appropriate conditions. Could gluten thus be converted into a more easily digestible form for humans?

Gluten is broken down naturally during the germination of cereal grains and by bacteria in the intestine and during the fermentation of sourdoughs.

To supply the germ with amino acids, the grain kernel synthetizes enzymes (proteases) that break down gluten proteins (Kucek et al. 2015). In order to use this ability of the cereal enzymes for human nutrition, cereal grains must germinate (Hartmann et al. 2006). The germ buds can then be processed into flour or bran and added during dough production. Some of the cereal enzymes work particularly efficiently under acidic conditions so that gluten breakdown is more effective in sourdoughs than in non-acidified doughs (Loponen et al. 2007). The cereal enzymes are also effective in the stomach and could break down gluten before it reaches the intestine.

For the preparation of sourdough, lactobacilli are typically added to a flour-water mixture. When incubated for a few hours at room temperature, the lactobacilli that break down gluten proteins multiply in the dough. Lactobacilli produce acid which not only gives the bread a special taste and consistency, but also makes it easier to digest due to the partially degraded gluten protein. Similar to the lactic acid bacteria used in the preparation of sourdough, bacteria in our intestine can also break down gluten peptides. For this reason, the quality of our microbiota, that is, the entity of the bacteria in our intestine, plays an important role in gluten sensitivity.

Why there is (almost) no Gluten in Malt
Malt lemonades, like Bionade®, are gluten free, although they contain barley. This is due to the malting process. During malting, grains kernels are soaked in water and germinated. This activates enzymes that breakdown starch and proteins stored in the kernel. Smaller gluten peptides are formed from gluten proteins, from which amino acids are finally released, which the germ bud needs for growth. Interestingly, certain cereal enzymes, known as endoproteases, cleave toxic gluten peptides next to the amino acids proline (P) or glutamine (Q). The potentially dangerous sequence PQLPY, which is present in several copies in the 33mer, is thus degraded during malting (Schwalb et al. 2012).

2.4 Non-Gluten Proteins

The non-gluten proteins of wheat make up 15–20 % of the total protein (Schalk et al. 2017b) and have important functions in the plant's metabolism and in its defense against parasites. In connection with wheat sensitivity, amylase-trypsin inhibitors (ATIs) and lipid transfer proteins are discussed (Juhasz et al. 2018). These non-gluten proteins, like some gluten proteins, belong to the protein family of prolamins, but in contrast to gluten proteins they can be extracted from wheat dough with saline solution. Since ATIs in combination with gluten appear to play a major role in the development of wheat sensitivity, these proteins will be discussed in more detail below.

ATIs account for up to 4 % of wheat protein and exist in different variants (Geisslitz et al. 2018). The active form typically consists of a combination of different ATI proteins, so-called subunits (Altenbach et al. 2011). Most of the ATI proteins known so far are encoded in the B and D genome, so that tetra- and hexaploid wheat species, such as emmer, spelt, durum, and bread wheat contain more ATIs than diploid wheat species, such as einkorn, which only contain the A genome. In some studies, no ATI was detected in einkorn (Geisslitz et al. 2018). However, ATI is also found in soy, buckwheat, millet, and other plants.

The function of the ATIs is not precisely known. Most likely, different combinations of subunits have different functions. As the name implies, ATIs are inhibitors of amylase and trypsin. Amylase is an enzyme that breaks down the plant starch amylose, a polymeric carbohydrate made up of glucose units. Trypsin is a digestive enzyme that is produced in the pancreas in humans and is responsible for the digestion of food proteins in the intestine. Trypsin-like proteins, called proteases, are found in plants and other organisms which breakdown the storage proteins in the wheat kernel and thus ensure the energy supply of the seedling. ATIs could play a role in the growth of the seedling by preventing storage carbohydrates and storage proteins from being broken down prematurely. Only when the grain begins to germinate the activity of ATIs declines, and the energy suppliers glucose and amino acids are released from the energy stores. There are also indications that ATIs inhibits enzymes in the digestive tract of insects, for example mealworms, but also of mammals.

When ATI proteins are broken down, which can happen in both cereal grains and the human intestine, valuable amino acids are produced that can be used as energy suppliers or for the biosynthesis of new proteins. In contrast to gluten proteins, ATIs are not particularly rich in the amino acids proline or glutamine. However, ATIs can also resist digestion in the intestine and trigger an immune response in sensitive individuals. There is evidence, particularly from experiments with mice, that ATIs associated with gluten have a damaging, pro-inflammatory effect. ATIs from soy, buckwheat or rice and other starchy plants, seem to have no or only weak immunogenic properties (Zevallos et al. 2017). It is postulated that the immunogenic effect of ATIs is based on an activation of innate immunity (Schuppan and Zevallos 2015). ATIs bind to a pattern recognition receptor on the surface of cells of the innate immune system, e.g. macrophages, monocytes, and dendritic cells, and stimulate these cells to release pro-inflammatory molecules (cytokines) (Fig. 4.3).

How can one find out whether gluten or ATI proteins have adverse health effects? Since gluten is present in a wheat grain in much larger quantities than ATIs, it is difficult to distinguish the toxicity of ATIs from that of gluten. The fact that ATIs can be harmful to health comes from patients with a disorder called non-celiac gluten sensitivity (NCGS), in whom gluten-induced celiac disease could be medically ruled out (Chap. 4). There is evidence that in these patients ATI peptides in the intestine trigger an inflammatory reaction (Schuppan and Zevallos 2015). An immune reaction occurs primarily in an already damaged intestinal epithelium. There are also indications that certain subunits of ATIs can trigger allergic reactions, especially baker's asthma (Juhasz et al. 2018). In healthy people, however, the small amounts of ATIs in the diet do not have any negative effects.

Interestingly, the noxious effect of ATIs could be reduced by lactobacilli, as found in a healthy intestine or used in the production of sourdough (Caminero et al. 2019). These bacteria contain ATI-degrading enzymes.that the negative effect of ATIs depends on the type of cereal and on the individual person can therefore be explained, among other things, by the methods of cereal processing (e.g., the production of a slowly fermenting sourdough) and the variability of the intestinal microbiota.

ATIs
These are found in various plants. They can trigger an inflammatory reaction in the intestine or allergic asthma in sensitive individuals.

ATIs can be broken down by bacteria—in the intestine or during the preparation of sourdough.

2.5 FODMAP

Carbohydrates make up the largest proportion in a ripe wheat grain kernel, at around 60%. These consist of about 90 % digestible starch, amylose and amylopectin, which is made up of long, partially branched chains of up to 100,000 glucose molecules. The remaining approximately 10 % are indigestible carbohydrates, which make up the dietary fiber content of cereals and are heterogeneous in size and composition. Among the dietary fibers, the so-called FODMAP play a special role (Biesiekierski et al. 2011). In wheat, FODMAP make up to 2 % of the carbohydrates and consist mainly of fructans, short-chain sugar molecules with a high fructose content. Fructans serve the plant as reserve carbohydrates and play a role in growth and stress resistance. Genes involved in the FODMAP metabolism of wheat are found in all types of wheat—from einkorn to bread wheat. However, other plants, such as corn, pulses, onions, or apples, are also rich in FODMAP (Varney et al. 2017).

In human nutrition, FODMAP plays an important role as prebiotics, as they can be utilized by intestinal bacteria, especially in the large intestine, and thus influence the intestinal microbiota and intestinal health. With a balanced intestinal flora and an intact intestinal epithelium, FODMAP may have anti-inflammatory properties. However, if the microbial milieu is disturbed (e.g., by nutrition, stress, infections) or if very large quantities are consumed, FODMAP can accumulate in the large intestine and be broken down by bacteria producing gases like hydrogen, carbon dioxide, and methane.. The consequences are flatulence, diarrhea and abdominal pain, but also changes in metabolism and impaired communication between the intestine and other organs (e.g., brain) with effects on the entire organism (De Giorgio et al. 2016). In persons with non-celiac non-wheat allergy-wheat sensitivity, FODMAP is discussed as symptom-promoting. Patients with irritable bowel syndrome can poorly tolerate foods containing FODMAP. It is unclear whether the effect is directly based on the metabolism of FODMAP or indirectly caused by an imbalance of the intestinal microbiota.

FODMAP
Fermentable oligo-, di-, and monosaccharides and polyols, make up the indigestible dietary fiber in wheat and other plants. In a healthy intestine, FODMAP serve as prebiotics and have an anti-inflammatory effect. FODMAP may worsen symptoms in individuals with inflammatory bowel disease and non-celiac non-wheat allergy wheat sensitivity,

FODMAP is found in comparable quantities in all wheat species. However, the FODMAP content is largely determined by the method of grain or flour processing.. The FODMAP content can be reduced by up to 90 % by proofing a dough made of bread wheat and baker's yeast for 4 h (Ziegler et al. 2016).

The Role of Intestinal Health

The intestine is our largest digestive organ, our largest immune organ, and an important part of our nervous system. Damage to the intestine can, therefore, lead to extensive loss of function, which not only affects digestion, but can also be accompanied by inflammatory reactions and mental illness (Shashikanth et al. 2017).

In order to be able to fulfill its many different tasks, the intestine has a special anatomical structure and is equipped with many different cells.

3.1   The Intestine as a Digestive Organ

The longest section of the intestine is the small intestine with a length of about 5m. It plays a central role in the digestion of food components and the absorption of nutrients. It is divided into the sections duodenum, jejunum, and ileum, with the duodenum following the stomach and the ileum being followed by the large intestine. The surface area of the small intestine is enormously increased by a complex folding system (Helander and Fandriks 2014). First, the intestinal mucosa is folded. These large folds are laid into small micro folds, the villi, which consist of villi and crypts. The villi are covered with epithelial cells (enterocytes), which are equipped with microvilli on the side facing the intestinal lumen. This approximately 100-fold increase in the surface area ensures efficient absorption of nutrients and secretion of digestive enzymes at the top of the villi. At the base of the crypts, there are stem cells from which the epithelium can regenerate (Fig. 3.1).

The colon, the upper part of the approximately 1.5 m long large intestine, is responsible for the utilization of dietary fiber. The lower part of the large intestine, the rectum, is used to store stool and is followed by the anus, the opening at the

© Springer Fachmedien Wiesbaden GmbH, part of Springer Nature 2021
C. Harter, *Gluten Sensitivity,* Springer essentials,
https://doi.org/10.1007/978-3-658-32657-9_3

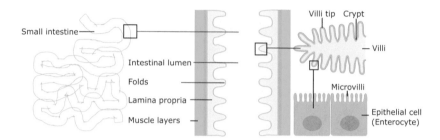

Fig. 3.1   Folding of the surface of the small intestine

end of the intestinal tract. Along its entire length, the surface of the intestine is covered with microbes, the intestinal microbiota, which comprises 10^{13}–10^{14} cells. The vast majority of the microbes are bacteria in the colon.

For the digestion of food components, the small intestine receives digestive enzymes from the pancreas. These enzymes break down a large amount of the macromolecules in the intestinal lumen that are taken in with food into fragments or even into individual building blocks. Some digestive enzymes are also produced by the intestine itself. For example, on the surface of the intestinal epithelium, the sugar-degrading enzymes lactase and saccharase are expressed, which split lactose (milk sugar) and saccharose (cane or beet sugar) in their constituents glucose and galactose in the case of lactose, or glucose and fructose in the case of saccharose. If a person suffers from lactose intolerance, the enzyme required to digest lactose, is inactive and the lactose is fermented by gas-producing bacteria in the large intestine causing abdominal pain and diarrhea. Other food components that cannot be degraded and absorbed in the small intestine, such as gluten peptides, FODMAP, and other dietary fibers, also enter the large intestine where they can be partially broken down and utilized by bacteria. Alternatively, substances that cannot be digested by human digestive enzymes may be excreted.

The digestible nutritional components, together with micronutrients such as vitamins and trace elements, are absorbed by the epithelial cells of the intestine on the apical side (facing the intestinal lumen) and are released into the blood (or in the case of lipids into the lymph) on the opposite, basal side (facing the blood) and delivered to the various organs.

To ensure that the intestine only absorbs substances that can be utilized, the epithelial cells on their surface are equipped with specific transport proteins. If no transporter is present, the substances normally remain in the intestinal lumen. The transport through the cells is called transcellular. The passage of food components or microorganisms between the cells is prevented in a healthy intestine by tight junctions. These consist of different proteins which, like closing strips, seal the space between neighboring epithelial cells. The integrity of the tight junctions is particularly important, as it is disturbed in many inflammatory bowel diseases. In a healthy state, only very small molecules, such as water, can pass between the epithelial cells. This transport is known as paracellular (Fig. 3.2).

Maldigestion and Malabsorption
Insufficient breakdown of food components due to a lack of or poorly working digestive enzymes or bile acids is known as maldigestion. The reasons may be insufficiencyof the pancreas, which produces the majority of digestive enzymes, or damage of epithelial cells of the small intestine that produce enzymes to break

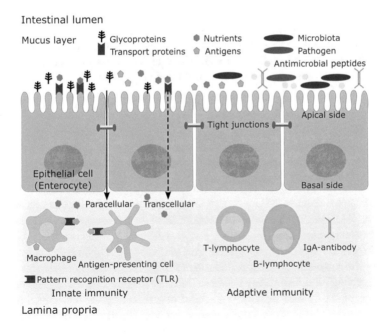

Fig. 3.2 Barriers and protective mechanisms of the intestine

down sugar molecules or small peptides. A lack of bile acids contributes to the maldigestion of fats.

In case of malabsorption, the degraded food components as well as vitamins and minerals supplied with the food are insufficiently absorbed by the epithelial cells of the intestinal mucosa. The reasons may be damage to the intestinal mucosa due to destruction of the villi, a so-called villous atrophy, or inflammation.

The consequences of maldigestion and malabsorption are a deficiency in the body's supply of nutrients and vitamins, but also diarrhea because the food components remain in the intestine and draw water from the tissues.

3.2   The Intestine as an Immune Organ

As an immune organ, the intestine protects us from unwanted intruders that enter our body with food. These can be pathogenic microorganisms that we ingest with food or food ingredients that can harm us.

In order to ensure that only harmless, usable substances enter our organism, the intestine has various mechanisms. These include the formation of physical barriers and the activation of the immune system (Fig. 3.2).

The Most Important Protective Mechanisms of the Intestine Include

- A mucus layer of glycoproteins on the surface of the epithelium that lines the luminal side. It prevents the penetration of pathogenic microorganisms by trapping them. In the colon, a second, mobile layer is added, which provides a habitat for nonpathogenic microorganisms and excludes the pathogenic ones.
- A complex immune system,—gastrointestinal mucosa-associated lymphatic tissue (GALT)—which consists of components of innate, nonspecific, and adapted, specific immunity.
- Cells of the innate immune system react very quickly to undesirable substances by secreting antimicrobial peptides that kill pathogenic microorganisms or by triggering an inflammatory response. In order to recognize the undesired substances, the immune cells carry so-called pattern recognition receptors (toll-like receptor, TLR) on their surface with which they identify "enemies" by means of specific signatures. Immune cells of the adaptive immune system react specifically to undesired substances by ensuring that molecules specifically directed against the intruder are produced. A typical reaction of the adaptive immune system is the production of antibodies against certain antigens. The antigens can

be derived from food or be structures of microbes. If the antigen penetrates the intestinal mucosa, antigen-presenting cells, macrophages, T and B lymphocytes are activated in the underlying connective tissue layer, the lamina propria. Finally, class A antibodies, IgA, which are typical for the intestinal immune system, are produced and transported into the intestinal lumen. There they recognize and inactivate the antigen. A specific, adaptive immune response requires several activation steps and takes days to weeks before it becomes effective.

Immunological Tolerance   Normally, the intestine does not direct a defensive reaction against molecules that we take in with food or against the "good" intestinal bacteria. This property of the intestinal immune system is called tolerance. Our organism learns immune tolerance from birth onwards. As soon as the protection provided by the maternal immune system declines, the infant must develop its own immune system and deal with the foreign molecules coming from outside. The more diverse the external world presents itself, the better the chances are of developing tolerance against as many molecules as possible. For this reason, it is advisable to confront infants from the 4th month of life with as many different foods and microorganisms as possible. This can drastically reduce the risk for allergy or food intolerance.

In order to avoid gluten sensitivity, it is recommended that children from the 4th month of life, but before the 12th month of life receive a gluten-containing diet for the first time.

The principle of tolerance development is also the basis of hyposensitization as a treatment of allergies: Over a period of several years, the substance to which the person is allergic, the allergen, is applied in increasing concentrations, for example, in the form of tablets. The organism thereby learns to tolerate the substance and in case of a response to the therapy, the allergic reaction disappears or at least becomes weaker.

3.3   The Microbiota of the Intestine

Several billion microorganisms inhabit our intestines and have a significant influence on our health. They influence not only the digestion of food and the utilization of nutrients but also our susceptibility to infections, the development of our immune system, and our behavior (Cryan and Dinan 2012; O'Connor 2013).

The entity of intestinal microorganisms, the microbiota, consists of 10^{13}–10^{14} cells. The human organism contains—without its microorganisms—"only" about

4×10^{13} cells. The microbiota weighs up to 2 kg and consists mainly of bacteria and to a small extent of fungi, viruses, and other single-cell organisms. The frequently used term "microbiome" refers to the entire genetic material of the microbes. It is estimated that the intestinal microbial genome contains about 3 million protein-coding genes, 150 times more than the human genome (Qin et al. 2010).

The intestinal microbiota in all humans contains a similar composition with respect to the bacterial strains (phyla), but there is great variability with respect to the members of a strain (species) (Zeißig 2016). The development of the microbiota of an individual is determined by the microbiota of the mother, genetic factors of the family, and environmental conditions, such as nutrition or stress. It is ultimately the result of a symbiotic relationship: Humans provide the bacteria a suitable habitat and the bacteria help humans digest, for example, by breaking down plant carbohydrates and producing valuable metabolic products for them. The microbiota of a healthy intestine also protects against entering of harmful substances into the body and colonization with pathogenic microorganisms as well as excessive immune reactions.

Important Functions of the Intestinal Microbiota are

- Support for the utilization of food components that are indigestible for humans
- Production of metabolites for the nutrition of intestinal cells (enterocytes) and thus maintenance of an intact intestinal epithelium
- Production of metabolites for the nutrition of health-promoting intestinal bacteria and thus maintenance of a balanced microbiota
- Regulation of gene expression, for example, to inhibit the growth of cancer cells in the colon
- Maintenance and modulation of the innate and adaptive immune system, for example, through anti-inflammatory properties
- Synthesis of mediators which communicate with other organs and thereby influence behavior

What Characterizes a "Healthy" Microbiota? In a healthy microbiota, the bacteria work hand in hand (Harmsen and de Goffau 2016). Some bacteria not only produce valuable substances for humans, but also for neighboring bacteria. For this purpose, the bacteria utilize suitable nutrients from human food, called prebiotics. Prebiotics include indigestible dietary fibers such as FODMAP and other oligosaccharides found in the hull of cereal grains, in pulses, onions, apples, and other fruits and vegetables. By breaking down the prebiotics, especially into short-chain carboxylic acids (butyrate, propionate, and acetate), the healthy microbiota is maintained (Louis et al. 2016).

Some species of Clostridia (Group IV and XIVa), for example, which produce butyrate and other short-chain fatty acids, are particularly useful bacteria (but beware: *Clostridium difficile* is a particularly harmful bacterium). Probiotics such as *Bifidobacterium* and *Lactobacillus* also have a beneficial effect by producing propionate and acetate and, among other things, inhibiting the formation of certain messenger substances that are involved in the development of depression and other mental illnesses (Fung et al. 2017). *Faecalibacterium prausnitzii*, which occurs mainly in the large intestine, is said to have a positive effect on our immune system.

How can the Microbiota be Disturbed?
Adisturbed ecology in the intestine, a dysbiosis, is a breeding ground for numerous diseases. It is, therefore, not surprising that in individuals with chronic inflammatory bowel diseases, an imbalance in the composition of the intestinal microbiota in favor of "bad bacteria" can be detected (Matthes 2016). It is unclear whether the unfavorable microbiota leads to the disease or whether other factors, for example, genetic disposition or nutrition, lead to damage of the intestinal mucosa and finally to dysbiosis. It is certain that an imbalance of the microbiota promotes inflammatory reactions and leads to a leakage of the intestinal epithelium. Once the microbial ecosystem is out of balance and the epithelium is damaged, a vicious circle begins that maintains inflammatory reactions and is difficult to interrupt. A remedy may be therapies with "good" bacteria, probiotics that can be administered orally, or stool transplants from healthy volunteers (Kump and Högenauer 2016). However, in most cases, genetic factors also play a role in the development of chronic inflammatory diseases and cannot be eliminated.

The microbiota of patients with celiac disease also contains less of the "good" bifidobacteria and lactobacilli, but more of the "bad" proteo- and enterobacteria (Verdu et al. 2015). However, it is unclear whether this is due to the gluten-free diet or to the genetic disposition of the patients. Similar to patients with chronic inflammatory bowel diseases, the intestinal epithelium of patients with celiac disease can no longer fulfill its barrier function. Since celiac disease is highly correlated with genetic factors, a change in the microbiota alone will not be able to provide a cure. Rather, the amount of gluten peptides (Sect. 2.3) must be reduced and the immune response must be suppressed (Sect. 4.1).

Disorders Associated with Gluten Sensitivity

The incidence of wheat-dependent diseases in Germany is estimated by experts to be up to 7% (Felber et al. 2014). A distinction is made between antigen-specific and antigen-unspecific diseases (Fig. 4.1). Antigen-specific diseases are triggered by an immune reaction towards a specific substance, the antigen. Only in very few cases it is certain that gluten is the antigen, that is, the disease-causing agent. In many cases, other ingredients of wheat seem to trigger or aggravate the disease. Often the disease-causing agent can only be speculated about. Celiac disease is clearly attributable to gluten, while some allergies are attributable to wheat. When neither celiac disease nor a wheat allergy can be diagnosed, the disorder is called"non-celiac non-wheat allergy wheat sensitivity. In common parlance, the term "gluten sensitivity" is often used to refer to "wheat sensitivity," although there are more factors that can be considered for the latter than for the former.

4.1   Celiac Disease

In Germany, the prevalence of celiac disease is 0.3–0.7% (Koletzko 2013), although the number of patients has increased in recent years. In other European countries the prevalence is higher, particularly in Sweden and Finland. On average, about 1% of the population is affected. It is unclear whether the increase in the disease is due to environmental factors, such as changes in diet, increased intestinal infections, and early use of antibiotics already in childhood, or to better diagnostic methods or changes in awareness (Lebwohl et al. 2018). Indeed, the age of diagnosis has shifted from early childhood to school age and in some cases to adulthood.

© Springer Fachmedien Wiesbaden GmbH, part of Springer Nature 2021   25
C. Harter, *Gluten Sensitivity,* Springer essentials,
https://doi.org/10.1007/978-3-658-32657-9_4

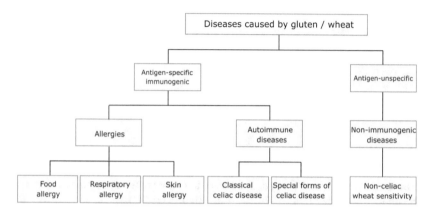

Fig. 4.1   Diseases caused by gluten or wheat

Celiac disease is an autoimmune disease that is triggered in genetically predisposed individuals by an immune reaction against gluten and persists throughout life. In its classic form, celiac disease is a chronic inflammatory bowel disease with malabsorption and abdominal discomfort. However, the clinical appearance is very complex and heterogeneous. Apart from the small intestine as the primarily affected organ, organs outside the digestive tract may also be affected. Patients with celiac disease sometimes suffer from depression, fatigue, and insomnia; the liver, bones, or nervous system may also be affected (Felber et al. 2014).

Special Forms of Celiac Disease Are

- Dermatitis herpetiformis, a chronic inflammatory, blistering skin disease
- Cerebral ataxia, which is associated with seizures and motor disorders
- Refractory celiac disease, in which patients do not respond to a gluten-free diet
- Potential celiac disease, in which patients are symptom-free despite genetic and serological indications.

Celiac disease can be clearly diagnosed. Irrespective of the clinical manifestations, antibodies against tissue transglutaminase (tTG) can be detected in the blood of patients. Since tTG is a protein produced naturally in the body and only patients with celiac disease produce antibodies against it, that is, recognize tTG as foreign, celiac disease is considered an autoimmune disease.

Only individuals with a genetic predisposition may suffer from celiac disease. Approximately 85–90% of those affected by celiac disease in Europe carry the histocompatibility marker HLA-DQ2, about 10–15% carry the marker HLA-DQ8. If these characteristics are missing, celiac disease can be practically ruled out. Since 25–35% of the population is positive for one of these characteristics, but less than 1% of the population in Germany has celiac disease, there must be other, previously unknown factors that contribute to the clinical picture.

Histologically, patients with celiac disease (with the exception of individuals with potential celiac disease) show changes in the mucous membrane of the small intestine, which are divided into different stages according to their severity and typically exhibit the following characteristics:

- An increased number of intraepithelial lymphocytes (IEL). These cells indicate an inflammatory reaction. The higher the number of IEL, the stronger the inflammatory response and the greater the damage to the intestinal mucosa.
- Prolongation of crypts (crypt hyperplasia) as a sign of remodelling of the intestinal mucosa due to inflammatory processes.
- Shortening to complete flattening of the villi, a so-called villous atrophy. The atrophy of the villi leads to disturbed absorption of nutrients from the intestine (malabsorption) and chronic inflammation.

The destruction of the intestinal mucosa is reversible. By removing gluten from the diet, the intestinal mucosa can regenerate completely, so that patients with celiac disease can live symptom-free.

Characteristics of Celiac Disease

- **Intestinal symptoms*:** Abdominal pain, diarrhea, constipation, flatulence, nausea, loss of appetite
- **Extraintestinal symptoms*:** Anemia, depression, insomnia, fatigue, weight loss, osteoporosis, etc.
- **Histology of the small intestine:** Increased number of IEL, villous atrophy
- **Genetic traits:** HLA-DQ2 or HLA-DQ8
- **Serum traits:** Antibodies against tissue transglutaminase, possibly antibodies against endomysium, antigliadin antibodies
- **Therapy:** Strictly gluten-free diet

*The symptoms listed represent a selection that may occur in different forms or may be absent. In any case, medical advice is required to make a diagnosis.

How can gluten ingested with food trigger celiac disease? The postulated pathome-chanism includes the following steps (Sollid et al. 2015) (Fig. 4.2):

- In the intestinal lumen, immunogenic gluten peptides are produced from larger gluten peptides by proteolytic cleavage. These gluten peptides are either trans- or paracellularly transported into the tissue layer below the intestinal epithelium, the lamina propria. The exact mechanisms by which gluten peptides enter the lamina propria are not known. It is only known that the epithelium of the small

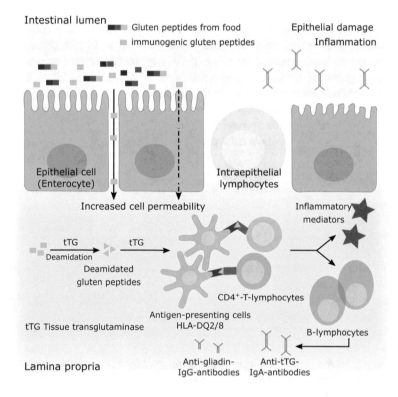

Fig. 4.2 Pathomechanism of celiac disease

intestine of patients with celiac disease is permeable to gluten peptides, whereas in healthy individuals the gluten peptides remain in the intestinal lumen.

- In the lamina propria, tTG converts some of the glutamines contained in gluten peptides, into glutamates. This process is called deamidation. It is a prerequisite for triggering a specific immune response.
- Deamidated gluten peptides are bound on the surface of certain antigen-presenting cells (APC-HLA-DQ2 or APC-HLA-DQ8).
- Gluten peptide-bearing APC are recognized by a specific T-cell type, $CD4^+$-T-lymphocytes.
- $CD4^+$-T-lymphocytes trigger immune reactions in two directions:
  - They produce inflammatory mediators and thus trigger an inflammatory response which, among other things, leads to damage to the intestinal epithelium.
  - They activate B-lymphocytes that produce antibodies against tTG and gluten peptides. These antibodies (anti-tTG-IgA, anti-gliadin-IgG) are used to diagnose celiac disease.

4.2   Wheat Allergies

In the database of allergen families of the University of Vienna, 34 allergens are listed when entering "wheat," which can be assigned to more than 300 genes of bread wheat (Juhasz et al. 2018). Thus, bread wheat is one of the most allergen-rich foods. However, not every potential allergen causes an allergy and only relatively few people fall ill.

In general, allergens are nonpathogenic, environmental substances that only trigger an immune response in hypersensitive persons. The consumer must be informed about allergens in food (EU Regulation No. 1169/2011).

Data on the prevalence of wheat allergy vary regionally and according to age group between 0.6 and 3%, worldwide the prevalence is estimated to be below 1% (Zuidmeer et al. 2008). Especially in childhood, allergies can disappear by themselves or be replaced by an allergy to another substance from the environment.

Allergies—as well as celiac disease—are accompanied by inflammation and in most cases by the formation of antibodies, in this case class E immunoglobulins (IgE). Depending on the point of entry of the allergen into the body, an allergy manifests itself with different symptoms (Christensen et al. 2014). Typical are respiratory diseases, swelling of mucous membranes, or skin diseases; gastrointestinal symptoms may also occur. Common to the various forms of allergy is a very rapid physical reaction immediately after contact with the allergen. The following forms of allergy **caused** by **wheat** are distinguished

- **Food allergies** that occur after ingestion of the allergen via the mouth.
- A special form is the wheat allergy that only occurs after physical exercise, also known as **wheat-dependent exercise-induced anaphylaxis.**
- **Baker's asthma,** which occurs after inhalation of flour and is mainly a disease of the respiratory tract.
- **Urticaria** or **dermatitis,** inflammatory rashes caused by skin contact with wheat proteins

Diagnostic criteria for a wheat allergy are an inflammatory reaction of the skin, a so-called skin prick test, when wheat proteins are applied, and/or the detection of IgE antibodies in blood serum.

The triggering allergens can be different depending on the type of allergy. In clinical practice, the molecular identity of the antigen is unknown in most cases. However, ω-gliadins are considered to be the main allergens in food allergy caused by the consumption of wheat. In exercise-induced anaphylaxis, which only occurs in adults after consumption of wheat-containing products in connection with physical exercise under aerobic conditions, ω5-gliadin could be specifically identified as the allergen. Certain ATI proteins (CM16) are associated with the development of asthma in bakers and millers who are constantly exposed to flour.

Characteristics of Wheat Allergy

- Typical **allergic symptoms** such as swelling of mucous membranes, rhinitis, breathing difficulties, rashes
- Allergic symptoms appear **shortly after consumption** of wheat (within a few minutes to 2h)
- **Intestinal** and **extraintestinal symptoms** similar to those of celiac disease possible, often respiratory and skin manifestations
- **Small intestine histology:** No or little damage to the intestinal epithelium
- Exclusion of celiac disease
- **Skin prick test:** Positive reactions with wheat extract or specific wheat proteins
- **Serum traits:** Anti-wheat IgE, or IgE against specific wheat proteins
- **Therapy:** Wheat-free diet, avoid contact with wheat flour, avoidance of cosmetics containing wheat

4.3   Non-Celiac Non-Wheat Allergy Gluten Sensitivity

A connection between the consumption of gluten and gastrointestinal symptoms in the absence of a diagnosis of celiac disease or allergy was first described in the late 1970s (Ellis and Linaker 1978). However, it is only in recent years that gluten sensitivity has become a worldwide phenomenon (Sapone et al. 2012). Frequently, those affected define gluten as the pathogenic agent without medical diagnostic criteria being met, such as anti-tTG antibodies in the case of celiac disease or anti-wheat IgE antibodies in the case of a wheat allergy. In these cases without a medical diagnosis, the nomenclature non-celiac non-wheat allergy wheat sensitivity or—in short form—non-celiac wheat sensitivity (NCWS) or non-celiac gluten sensitivity (NCGS) has been coined.

Validated information on the frequency of NCWS/NCGS is not possible due to the lack of diagnostic criteria. The proportion of self-diagnosed patients is high, and it is difficult to distinguish between disease and fashion trends. The extent to which NCWS/NCGS is an independent clinical entity is controversially discussed (Reese et al. 2018).

A current description for NCWS/NCGS is: "A disease of multifactorial etiology due to a functional effect caused by FODMAP combined with a mild gluten-triggered immune response and microbial imbalance" (Plaum 2019). This report also mentions the involvement of ATI as activators of the innate immune defense.

Characteristics of NCWS/NCGS

- **Intestinal** and **extraintestinal symptoms** similar to those of celiac disease after consumption of foods containing wheat
- Symptoms appear **hours** or **days after consumption of wheat** and disappear after discontinuing foods containing wheat
- **Small intestine histology:** No or minimal damage to the intestinal epithelium
- Exclusion of celiac disease
- Exclusion of a wheat allergy
- **Therapy:** Nutritional advice and change of diet, if necessary wheat- and/or FODAMP-free diet

The diagnosis of NCWS/NCGS is currently made on the basis of exclusion criteria and the information provided by the person affected. Proof that gluten or wheat contribute to the clinical picture can only be provided by a controlled nutrition program. The gold standard test is a blinded, placebo-controlled, medically supervised study in which the affected person consumes a gluten-free or gluten-containing diet (without knowing whether he or she is on a gluten-free or gluten-containing diet). The patient's symptoms and assessment and, if necessary, clinical parameters are documented. The criterion for wheat/gluten sensitivity is an improvement of main symptoms of at least 30% after a 6-week gluten-free diet identified by the patient on a rating scale (Catassi et al. 2015). In order to secure the diagnosis, the test should be repeated after a waiting period of several weeks.

An evaluation of several studies showed that gluten as the trigger of the disease could not be confirmed in more than 80% of suspected NCGS patients (Molina-Infante and Carroccio 2017). About 40% of NCGS patients showed a nocebo effect, that is, the symptoms in the placebo group worsened or remained unchanged compared to the group that ingested the allegedly harmful gluten. The fact that the mere assumption of ingesting a harmful substance can trigger a symptom is a common phenomenon in food intolerances and makes it impossible to identify a trigger. Only in about 20% of self-diagnosed NCGS patients, gluten could be confirmed as a trigger for NCGS (Dale et al. 2018). In another study with self-diagnosed NCGS patients, it was shown that the symptoms improved with a FODMAP-free diet (Biesiekierski et al. 2013).

Due to the heterogeneous data situation, NCWS/NCGS can be regarded as a multifactorial phenomenon. There is no doubt that there are patients for whom a switch to a gluten-free diet has a symptom-relieving effect. A reduction of the amount of FODMAP in the diet can also lead to an improvement of the symptoms. Patients often notice for themselves which type of diet is good for them and choose the foods that are best tolerated. However, due to large nocebo effects, the "belief" in an "(un)healthy" food has a great influence on its tolerability. Ultimately, affected individuals may avoid a food without a confirmed diagnosis of intolerance.

Despite the lack of specific diagnostic criteria, there is a clinical picture of NCWS/NCGS. There are patients who complain of discomfort after consumption of wheat and in whom the symptoms improve after a change in diet. A possible explanation for the maintenance of the symptoms is a damaged intestinal epithelium, which is further damaged by the lifestyle and diet.

The following **factors** can contribute to the characteristics of **NCWS/NCGS**:

- Genetic/epigenetic predisposition (including HLA-DQ2/8)
- Pre- and postnatal nutrition (nutrition during pregnancy and lactation)
- Diet in adulthood (e.g., diet rich in saturated fatty acids, low-fiber diet, etc.)
- Imbalance of the intestinal microbiota (dysbiosis)

It probably requires a combination of different factors that lead to the symptoms of NCWS/NCGS. Once the intestinal epithelium is damaged and an inflammatory reaction has been initiated, a vicious circle is often maintained which can only be interrupted by a massive change in lifestyle.

The following **pathomechanisms** are discussed for the development of **NCWS/NCGS** (Potter et al. 2018; Schuppan et al. 2015) (Fig. 4.3)

- Increased permeability of the intestinal epithelium through dysbiosis/nutrition
- Gluten and/or ATI destroy the tight junctions between the epithelial cells and can enter the lamina propria in a paracellular fashion.
- Further antigens from food or microorganisms enter the lamina propria

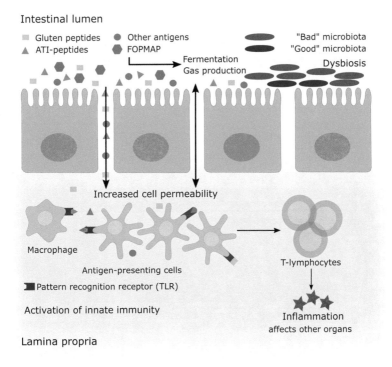

Fig. 4.3 Factors and mechanisms that can contribute to the development of NCGS/NCWS

- ATI activates the innate immune defence via pattern recognition receptors (TLR 4)
- ATI, gluten peptides and possibly other antigens trigger inflammatory reactions.
- FODMAP are increasingly degraded by gas-producing bacteria. The gases lead to flatulence, the degradation products influence the metabolism and the microbiota.

Gluten in the Diet

5.1 Advantages of a Gluten-Containing Diet

Gluten is not a modern invention. Cereals containing gluten have been part of the diet since mankind settled down and began to cultivate crops about 10,000 years ago. Since then, the composition of gluten has been optimized by breeding to improve the baking properties of the flours. Nevertheless, the proportion of gluten proteins in wheat grain has not changed dramatically in recent decades. In particular, modern bread wheat does not contain more immunogenic gliadin peptides than old varieties (Ribeiro and Nunes 2019). Nor have any foreign genes been introduced into wheat by manipulating the genetic material—in contrast to maize and soya. The human immune system has thus been able to deal with gluten as a food component over the past millennia and has generally developed oral tolerance to gluten.

After maize, wheat makes a major contribution to feeding the world population. Wheat is a nutritionally valuable food, especially when the whole grain is consumed.

The wheat grain kernel essentially consists of 3 parts (Hemery et al. 2007) (Fig. 5.1)

1. The epidermis and seed coat that envelop the kernel and contain most of the fiber and some of the minerals.
2. The endosperm, surrounded by the aleurone layer, which comprises about 85% of the kernel. The endosperm contains the grain's energy stores, carbohydrates (amylose and amylopectin), and storage proteins, mainly gluten. The aleurone layer contains the highest concentrations of minerals and water-soluble vitamins.

© Springer Fachmedien Wiesbaden GmbH, part of Springer Nature 2021
C. Harter, *Gluten Sensitivity,* Springer essentials,
https://doi.org/10.1007/978-3-658-32657-9_5

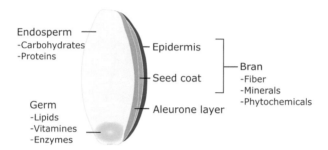

Fig. 5.1   Components of a wheat grain kernel

3. The seedling (germ) with cotyledon and cotyledon root, which comprises only about 3% of the kernel, but makes up the most nutritious part.. It contains not only proteins but also valuable lipids and fat-soluble vitamins, found in wheat-germ oil, for example.

When a wheat kernel is processed, the endosperm is always used. Thus, wheat products contain the carbohydrates and proteins stored in the endosperm. To produce different types of flour, the germ and, in varying proportions, the hull are separated. For example, 100 g of flour of type 405 contains 405 mg ash, which corresponds approximately to the mineral content. Flours with higher type numbers contain correspondingly higher proportion of fiber and minerals. In the production of wholemeal flours or wholemeal meals, the entire grain kernel, including the germ, is used after removing only a thin layer of the outer hull. This not only makes whole grain products particularly rich in fiber and minerals but also in vitamins and secondary plant compounds.

Cleaned and dried wheat kernels contain on average 65% longchain and complex carbohydrates, 15% protein, 13% fiber, 2% fat and various vitamins (especially vitamins B1, B6, E), minerals (especially iron, magnesium, zinc), as well as secondary plant compounds (especially polyphenols, phytosterols).

A diet with whole wheat products can easily cover the recommended fiber content of about 30g per day. Corn or rice, but also the pseudo-cereals amaranth and quinoa contain much less fiber. Wholemeal products made from wheat (or rye and barley) are also valuable in meeting the need for certain minerals and vitamins. Furthermore, the high content of complex carbohydrates and the low-fat content of wheat contribute to a healthy diet (Healthgrain Forum).

The average diet in the western industrialized nations contains too much fat and too much refined sugar, but too little fiber and complex carbohydrates. This rather unhealthy diet causes a number of lifestyle diseases such as obesity, diabetes mellitus type II, coronary heart disease, and dyslipidemia combined with high cholesterol levels. With a diet rich in whole grain cereal products, it is easy to eat a healthy diet without having to resort to food supplements. Only the proteins in cereals are not of high quality for human nutrition, hence the diet should be supplemented with proteins from animal products or legumes and pseudo-cereals.

Within the framework of the "Healthgrain Project," an EU-funded research project, the extent to which a diet with whole grain products affects the occurrence of typical chronic diseases of the western industrialized countries, so-called "lifestyle" diseases (Bjorck et al. 2012), was investigated.

In summary, the studies provided the following results (Hauner et al. 2012).

A Diet with Whole Grain Products

- Reduces the risk of obesity
- Reduces the risk of developing diabetes mellitus type II ("adult-onset diabetes" or diet-related diabetes)
- Reduces the risk of developing high blood pressure
- Reduces the risk of coronary heart disease
- Reduces plasma levels of total cholesterol and LDL cholesterol ("bad" cholesterol).
- Reduces the risk of colon cancer

What is the Reason for the Positive Effects of Whole Grain Cereals ?

The health-promoting properties are attributed above all to dietary fibers and secondary plant compounds (Bach Knudsen et al. 2017). Dietary fibers represent a heterogeneous group of food constituents whose structure and complexity is only partially known. However, all dietary fibers have in common that they contain poly- or oligomeric carbohydrates that cannot be degraded by the human digestive system and cannot be absorbed in the small intestine. They reach the large intestine, where the fermentable dietary fibers can be utilized by intestinal bacteria. The nonfermentable dietary fibers are excreted undigested or only slightly altered. This increases intestinal motility, shortens the transit time of food components in the intestine and increases stool volume, which has a positive effect on intestinal and metabolic functions. In addition, dietary fibers, as prebiotics, serve to maintain a healthy microbiota by promoting the proliferation of "good" intestinal bacteria and at the same time preventing the proliferation of "bad" intestinal bacteria.

Various bioactive substances are bound to dietary fibers, so-called phytochemicals or secondary plant compounds. These include flavonoids and aromatic carboxylic acids. They dissociate from the dietary fibers in the intestine and are partially modified by the intestinal bacteria, which can then be absorbed by the intestinal epithelial cells and distributed in the body via the blood. They are said to have anti-oxidative and anti-inflammatory properties.

Conclusion Gluten-Containing Diet
For healthy people, a gluten-containing diet has positive effects for several reasons: It contains many nutritionally valuable ingredients and it supports the maintenance of a healthy intestinal microbiota enriched with useful, anti-inflammatory bacteria.

Gluten is inseparably linked to the wheat grain. While the fibers, minerals, and phytochemicals are mainly contained in the outer layers, the bran, gluten is found inside the grain, the endosperm. Bran and endosperm cannot be completely separated from each other. A fiber-rich diet with wheat, therefore, always contains gluten.

5.2   Advantages of a Gluten-Free Diet

Patients with diagnosed celiac disease or wheat allergy should eat a gluten-free diet. In these patients, gluten triggers inflammatory reactions that not only damage the intestinal epithelium and thus impair intestinal health, but can also have an effect on other tissues such as bones, liver or the nervous system. In the long term, a gluten-containing diet can lead to serious, often irreversible damage of the entire organism in these patients. However, a gluten-free diet can help most patients with celiac disease or wheat allergies to live free of symptoms.

In patients with NCWS/NCGS, a gluten-free diet can help to improve symptoms. However, in most cases gluten itself may not be the cause of the disorder. Rather, other dietary or environmental factors such as FODMAP, alcohol, high-fat diet or stress, may contribute to microbial dysbiosis, leakage of the intestinal epithelium, and inflammation (Dieterich et al. 2019). If intestinal health is impaired, the organism reacts much more sensitively to food components such as gluten or lactose, although it can tolerate these substances well with a healthy intestine.

It is often worth trying to find out the component that triggers the disease by consciously reducing, in addition to wheat products, for example, FODMAP or fat in the diet. Sometimes the intestinal epithelium and the intestinal ecosystem regenerate through a change in diet. After the symptoms have improved, small amounts of gluten or wheat products can be fed again.

Dietary fiber can also cause problems. Their positive properties as prebiotics are not useful for all individuals. Some people cannot tolerate dietary fiber well, because it promotes the growth of gas-producing bacteria that cause abdominal pressure and flatulence. In this condition, a reduction of dietary fiber in the diet can help..

Conclusion Gluten-Free Diet
For individuals with celiac disease, a gluten-free diet is currently the only effective treatment option. In the case of wheat allergies, the type of allergy determines whether a gluten-free diet can help. In NCWS/NCGS, gluten is rarely the disease-causing agent. In this case, the disease-causing factor should be identified—after medical consultation—by a targeted change in diet.

5.3   Disadvantages of a Gluten-Free Diet

The advantages of a gluten-containing diet are also the disadvantages of a gluten-free one. For example, a gluten-free diet may be deficient in complex carbohydrates and fiber as well as certain vitamins, minerals, and phytochemicals.

In order to avoid nutritional deficiencies, it is particularly important to pay attention to the quality of the food with which gluten is replaced (El Khoury et al. 2018). Replacement with corn or rice provides a diet of lower quality, which is naturally poor in fiber, vitamins, minerals, and phytochemicals. In an American study, even elevated concentrations of toxic arsenic and mercury were measured in the urine or blood of people who ate a gluten-free diet. Normally, no food should contain arsenic or mercury. It is suspected that rice flour was the cause of contamination, as rice is often grown in areas with arsenic and mercury-containing soils (Bulka et al. 2017). This dramatic example of a negative effect of a gluten-free diet underlines the importance of a balanced, high-quality diet. Buckwheat, amaranth, or quinoa can provide valuable grain substitutes. However, these pseudocereals can only partially compensate for the loss. On one hand, they are not grown to the extent that they could cover the global demand. On the other hand, they do not have the baking and sensory qualities to provide an adequate substitute for the usual baked goods and pasta.

In order to produce bakery products and pasta comparable to those containing gluten, various substances such as fibers, thickeners, or flavoring must be added to the gluten-free raw materials. The foods produced in this way often contain many more components than a product made from natural gluten-containing cereals, and often also synthetic fibers such as modified cellulose. The various additives carry the risk of causing intolerances.

Meta-analysis, in which several studies on gluten-free diets of patients with celiac disease were evaluated, have shown that there are deficits in the supply of dietary fiber, various vitamins, and minerals, while at the same time the content of fats and simple sugars is too high (Melini and Melini 2019).

The evaluation of the diet of several thousand people not suffering from celiac condition also provides evidence that a gluten-containing diet has positive effects on health compared to a gluten-free diet. A low-gluten content in the diet correlated with an increased risk of coronary heart disease and diabetes mellitus type II (Lebwohl et al. 2017). The authors of these studies conclude that the lower content of valuable dietary fibers in wheat grain in a low-gluten diet leads to an increased risk of disease and advise healthy people against a gluten-free diet.

Another disadvantage of gluten-free products is the high price: Foods that are typically made from gluten-containing cereals—such as baked goods and pasta—are often much more expensive in gluten-free variants than the corresponding gluten-containing ones.

Also, the belief that a gluten-free diet is healthier than a gluten-containing diet can lead to the development of an eating disorder, Orthorexia nervosa. People who are overly concerned about the quality of their diet are at risk. Another aspect concerns the nutrition of children: A gluten-free diet in childhood carries the risk of developing allergies because the organism does not have the possibility to develop oral tolerance to gluten.

Gluten-Free as a Trend

6

In a survey of 1,002 people (55% female, 45% male) in England in 2012, 13% said they suffer from gluten sensitivity, 79% of whom were female. However, only 3.7% of this group of people consumed a gluten-free diet and only 0.8% were diagnosed with celiac disease (Aziz et al. 2014). Similar results were obtained from surveys of other populations. In a study in New Zealand, it was found that out of almost 600 children about 5% avoid gluten, although only 1% were diagnosed with celiac disease (Tanpowpong et al. 2012). Among 910 athletes, including 18 world-class athletes, 41% stated that they follow a gluten-free diet (Lis et al. 2016). More than half of them diagnosed "gluten sensitivity" themselves, combined with the belief that avoiding gluten leads to better performance. However, in a blinded study, in which a part of the athletes received a gluten-free diet and the other part a gluten-containing diet, no effect of gluten on performance was found. The belief in the benefits of a gluten-free diet alone apparently drove the athletes to perform better. The authors of this study emphasize that athletes are particularly sensitive to gastrointestinal problems and eat a particularly health-conscious diet. Through their consumption behavior, they influence the market as well as consumers, although no data are available to justify their behavior on a scientific basis.

There are several statistical surveys on gluten-free consumer behavior in Germany. The most recent survey dates back to 2019: Of 1,047 people, 5% stated that they eat gluten free (Statista 2019). In a survey in 2015 of 554 consumers who deliberately bought organic food, 24% said they bought gluten-free food (Biopinio 2015). However, only about one in five people who bought gluten-free food consistently followed a gluten-free diet. The largest proportion of gluten-free shoppers occasionally ate gluten-free food. As reasons for consuming gluten-free food, almost 52% of 233 respondents said that they have the feeling that they cannot tolerate gluten well. Twenty-three percent ate gluten-free in order to lose weight, while only just under 19% stated that they had a medically proven gluten sensitivity (Statista 2017). According to a nationwide representative survey in 2016 (Techniker Krankenkasse; Technicians Health insurance company), 1% of Germans believe that they suffer from gluten sensitivity. This figure fits well with the medically proven figure, but is discrepant with consumer behavior.

In summary, based on the surveys, one can conclude that: There is a gap between the number of people who buy gluten-free food and the number of people who consume a gluten-free diet or suffer from medically proven gluten sensitivity.

The population's awareness of nutrition has changed in recent years toward healthier and more natural food. This rethinking is changing consumer behavior and thus the development of the market. The sales figures for organic food and free-from food (lactose-free, gluten-free) have been rising continuously for years. In 2017, the turnover of gluten-free food in Germany amounted to a good 170 million € (Lebensmittelzeitung 2017). Twenty-four percent of consumers, who already bought gluten-free products predominantly from organic food stores, found that the range of products on offer is not sufficient (Biopinio 2015). The food industry is responding promptly to consumers' wishes and has increased the proportion of newly introduced gluten-free products from 6 to 11% between 2011 and 2015 (Mintel 2016). That the food industry's investments could be worthwhile is shown by the assessment that 73% of 1000 respondents believe that gluten-free products will remain on the market in the long term, while only 16% believe that this is a fashion trend (Bundesministerium für Ernährung und Landwirtschaft 2016; Ministry of nutrition and agriculture).

Conclusion—Gluten-Free as a Trend

Of the consumers who buy gluten-free products, only a small proportion consistently eat a gluten-free diet.

As a reason for a gluten-free diet, the majority of those surveyed say that gluten is not good for them. Less than 20% eat a gluten-free diet on the basis of a medical diagnosis.

The majority of consumers believe that gluten-free products will remain on the market in the long term.

What You Learned From This *essential*

- Gluten is a complex mixture of different proteins found in wheat, barley, and rye.
- Bread wheat and spelt are genetically closely related and consist of 3 subgenomes, A, B, and D.
- Apart from gluten, amylase trypsin inhibitors (ATI) and fermentable oligo-, di- and monosaccharides and polyols (FODMAP) might be involved in wheat sensitivity.
- An undamaged intestinal epithelium and a balanced microbiota are prerequisites for wheat tolerance and a functional metabolism.
- Celiac disease is a gluten-induced disorder that only occurs in genetically predisposed individuals. It can be medically clearly diagnosed.
- Wheat allergies can be caused by gluten or other wheat proteins. They can present themselves as gastrointestinal diseases, respiratory, and skin diseases.
- Non-celiac non-wheat allergy wheat sensitivity represents a clinical picture with an unclear cause and various intestinal and extraintestinal symptoms. A combination of different factors—wheat ingredients, damaged intestinal epithelium, genetics, and environmental influences—may be the trigger.
- Cereals containing gluten are nutritionally valuable. A diet with whole-grain products containing gluten can reduce the risk of developing a lifestyle disease.
- Patients with diagnosed celiac disease or wheat allergy should adhere to a gluten-free diet.
- A gluten-free diet bears the risk of malnutrition with deficiencies in fiber and certain minerals and vitamins.
- About 5% of consumers in Germany consume gluten-free food, although only 1% suffer from a diagnosed gluten sensitivity. Most consumers believe that gluten-free foods will remain on the market in the long term.

© Springer Fachmedien Wiesbaden GmbH, part of Springer Nature 2021
C. Harter, *Gluten Sensitivity,* Springer essentials,
https://doi.org/10.1007/978-3-658-32657-9

References

Allergen Online. University of Nebraska-Lincoln. https://allergenonline.org. Accessed: 21 July 2019.

Altenbach, S. B., Vensel, W. H., & Dupont, F. M. (2011). The spectrum of low molecular weight alpha-amylase/protease inhibitor genes expressed in the US bread wheat cultivar Butte 86. *BMC Research Notes, 4,* 242.

Andersen, G., & Koehler, H., in collaboration wit Rubach, M., & Schaecke, W. (2015). Annual report of the German Research Institute (*Jahresbericht der deutschen Forschungsanstalt*) *2014* (S. 136–139). Freising. https://www.kern.bayern.de/mam/cms03/themen/bil der/flyer_gluten.pdf.

Aziz, I., Lewis, N. R., Hadjivassiliou, M., Winfield, S. N., Rugg, N., Kelsall, A., et al. (2014). A UK study assessing the population prevalence of self-reported gluten sensitivity and referral characteristics to secondary care. *European Journal of Gastroenterology and Hepatology, 26,* 33–39.

Bach Knudsen, K. E., Norskov, N. P., Bolvig, A. K., Hedemann, M. S., & Laerke, H. N. (2017). Dietary fibers and associated phytochemicals in cereals. *Molecular Nutrition & Food Research, 61,* 7.

Bickel, S. (2015). Unser tägliches Brotgetreide. *Biologie in Unserer Zeit, 45,* 168–175.

Biesiekierski, J. R., Rosella, O., Rose, R., Liels, K., Barrett, J. S., Shepherd, S. J., et al. (2011). Quantification of fructans, galacto-oligosacharides and other short-chain carbohydrates in processed grains and cereals. *Journal of Human Nutrition & Dietetics, 24,* 154–176.

Biesiekierski, J. R., Peters, S. L., Newnham, E. D., Rosella, O., Muir, J. G., & Gibson, P. R. (2013). No effects of gluten in patients with self-reported non-celiac gluten sensitivity after dietary reduction of fermentable, poorly absorbed, short-chain carbohydrates. *Gastroenterology, 145,* 320–328.

Biopinio. (2015). Ergebnisse zur Free-From Studie. https://biopinio.de/studie-free-from-2015/. Accessed: 19 July 2019.

Bjorck, I., Ostman, E., Kristensen, M., Anson, N. M., Price, R. K., Haenen, G. R. M. M., et al. (2012). Cereal grains for nutrition and health benefits: Overview of results from in vitro, animal and human studies in the HEALTHGRAIN project. *Trends in Food Science & Technology, 25,* 87–100.

Bulka, C. M., Davis, M. A., Karagas, M. R., Ahsan, H., & Argos, M. (2017). The unintended consequences of a Gluten-free diet. *Epidemiology, 28,* E24–E25.

Bundesministerium für Ernährung und Landwirtschaft. (2016). Wie schätzen Sie die folgenden Trend-Lebensmittel ein? Chart. 3. Januar 2017. Statista. https://de.statista.com/sta tistik/daten/studie/653684/umfrage/entwicklung-von-trend-lebensmittel-in-deutschland/. Accessed: 19 July 2019.

Bundesverband deutscher Pflanzenzüchter e. V. https://www.bdp-online.de/de/Pflanzenzuec htung/Kulturarten/Getreide/Weizen. Accessed: 14 July 2019.

Caminero, A., McCarville, J. L., Zevallos, V. F., Pigrau, M., Yu, X. B., Jury, J., et al. (2019). Lactobacilli degrade wheat amylase trypsin inhibitors to reduce intestinal dysfunction induced by immunogenic wheat proteins. *Gastroenterology, 156,* 2266–2280.

Catassi, C., Elli, L., Bonaz, B., Bouma, G., Carroccio, A., Castillejo, G., et al. (2015). Diagnosis of Non-Celiac Gluten Sensitivity (NCGS): The Salerno experts' criteria. *Nutrients, 7,* 4966–4977.

Christensen, M. J., Eller, E., & Bindslev-Jensen, C. (2014). Patterns of suspected wheat related allergy: A retrospective single-centre study of 156 patients. *Allergy, 69,* 257–257.

Codex Alimentarius Standard for foods for special dietary use for persons intolerant to gluten. Codex Stan 118–1979 Adopted 1979, Amendment 1983 and 2015, Revision 2008.

Cryan, J. F., & Dinan, T. G. (2012). Mind-altering microorganisms: The impact of the gut microbiota on brain and behaviour. *Nature Reviews Neuroscience, 13,* 701–712.

Dale, H. F., Hatlebakk, J. G., Hovdenak, N., Ystad, S. O., & Lied, G. A. (2018). The effect of a controlled gluten challenge in a group of patients with suspected non-coeliac gluten sensitivity: A randomized, double-blind placebo-controlled challenge. *Neurogastroenterology and Motility, 30,* e13332.

Dale, H. F., Biesiekierski, J. R., & Lied, G. A. (2019). Non-coeliac gluten sensitivity and the spectrum of gluten-related disorders: An updated overview. *Nutrition Research Reviews, 32,* 28–37.

Database of allergen families. www.meduniwien.ac.at/allfam. Accessed: 17 July 2019.

De Giorgio, R., Volta, U., & Gibson, P. R. (2016). Sensitivity to wheat, gluten and FODMAPs in IBS: Facts or fiction? *Gut, 65,* 169–178.

Dieterich, W., Schuppan, D., Schink, M., Schwappacher, R., Wirtz, S., Agaimy, A., et al. (2019). Influence of low FODMAP and gluten-free diets on disease activity and intestinal microbiota in patients with non-celiac gluten sensitivity. *Clinical Nutrition, 38,* 697–707.

Dvorak, J., Deal, K. R., Luo, M. C., You, F. M., von Borstel, K., & Dehghani, H. (2012). The origin of spelt and free-threshing hexaploid wheat. *Journal of Heredity, 103,* 426–441.

El Khoury, D., Balfour-Ducharme, S., & Joye, I. J. (2018). A review on the gluten-free diet: Technological and nutritional challenges. *Nutrients, 10,* 1410.

Ellis, A., & Linaker, B. D. (1978). Non-coeliac gluten sensitivity. *Lancet, 1,* 1358–1359.

EU Regulation Nr. 1169/2011. Article 21, Annex II.

EU Implementation Regulation Nr. 828/2014. Implementation Regulation on the requirements for the provision of information to consumers on the absence or reduced presence of gluten in food. Durchführungsverordnung über die Anforderungen an die Bereitstellung von Informationen für Verbraucher über das Nichtvorhandensein oder das reduzierte Vorhandensein von Gluten in Lebensmitteln.

Fasano, A. (2011). Zonulin and its regulation of intestinal barrier function: The biological door to inflammation, autoimmunity, and cancer. *Physiological Reviews, 91,* 151–175.

Felber, J., Aust, D., Baas, S., Bischoff, S., Blaker, H., Daum, S., et al. (2014). Results of a S2k-Consensus Conference of the German Society of Gastroenterolgy, Digestive- and

Metabolic Diseases (DGVS) in conjunction with the German Coeliac Society (DZG) regarding coeliac disease, wheat allergy and wheat sensitivity. *Zeitschrift fur Gastroenterologie, 52,* 711–743.

Ficco, D. B. M., Prandi, B., Amaretti, A., Anfelli, I., Leonardi, A., Raimondi, S., et al. (2019). Comparison of gluten peptides and potential prebiotic carbohydrates in old and modern Triticum turgidum ssp. Genotypes. *Food Research International, 120,* 568–576.

Forum Bio- und Gentechnologie e. V. https://www.transgen.de/datenbank/1995/weizen.html. Accessed: 14 July 2019.

Fung, T. C., Olson, C. A., & Hsiao, E. Y. (2017). Interactions between the microbiota, immune and nervous systems in health and disease. *Nature Neuroscience, 20,* 145–155.

Geisslitz, S., Ludwig, C., Scherf, K. A., & Koehler, P. (2018). Targeted LC-MS/MS reveals similar contents of alpha-Amylase/Trypsin-Inhibitors as putative triggers of nonceliac Gluten sensitivity in all wheat species except Einkorn. *Journal of Agriculture and Food Chemistry, 66,* 12395–12403.

Harmsen, J. M., & de Goffau, M. C. (2016). The human gut microbiota. In A. Schwiertz (Ed.), *Microbiota of the human body* (pp. 95–108). Cham: Springer.

Hartmann, G., Koehler, P., & Wieser, H. (2006). Rapid degradation of gliadin peptides toxic for coeliac disease patients by proteases from germinating cereals. *Journal of Cereal Science, 44,* 368–371.

Hauner, H., Bechthold, A., Boeing, H., Bronstrup, A., Buyken, A., Leschik-Bonnet, E., et al. (2012). Carbohydrate intake and prevention of nutrition-related diseases: Evidence-based guideline of the German Nutrition Society. *Deutsche Medizinische Wochenschrift, 137,* 389–393.

Healthgrain Forum. https://healthgrain.org/wp-content/uploads/2019/01/HGF_SS_Benefits-of-WG_2012-11-21.pdf. Accessed: 19 July 2019.

Helander, H. F., & Fandriks, L. (2014). Surface area of the digestive tract – Revisited. *Scandinavian Journal of Gastroenterology, 49,* 681–689.

Hemery, Y., Rouau, X., Lullien-Pellerin, V., Barron, C., & Abecassis, J. (2007). Dry processes to develop wheat fractions and products with enhanced nutritional quality. *Journal of Cereal Science, 46,* 327–347.

Herran, A. R., Perez-Andres, J., Caminero, A., Nistal, E., Vivas, S., Ruiz de Morales, J. M., & Casqueiro, J. (2017). Gluten-degrading bacteria are present in the human small intestine of healthy volunteers and celiac patients. *Research in Microbiology, 168,* 673–684.

International Wheat Genome Sequencing Consortium. (2018). Shifting the limits in wheat research and breeding using a fully annotated reference genome. *Science, 361,* eaar7191.

Juhasz, A., Belova, T., Florides, C. G., Maulis, C., Fischer, I., Gell, G., Birinyi, Z., Ong, J., Keeble-Gagnere, G., Maharajan, A., et al. (2018). Genome mapping of seed-borne allergens and immunoresponsive proteins in wheat. *Science Advances, 4,* eaar8602.

Koletzko, S. (2013). Diagnosis and treatment of celiac disease in children. *Monatsschrift Kinderheilkunde, 161,* 63–75.

Kucek, L. K., Veenstra, L. D., Amnuaycheewa, P., & Sorrells, M. E. (2015). A grounded guide to gluten: How modern genotypes and processing impact wheat sensitivity. *Comprehensive Reviews in Food Science and Food Safety, 14,* 285–302.

Kump, P., & Högenauer, C. (2016). Fäkale Mikrobiota-Transplantation. In A. Stallmach & M. J. G. T. Vehreschild (Eds.), *Mikrobiom* (pp. 299–319). Berlin: De Gruyter.

Lebensmittelzeitung. (2017). Umsatz mit glutenfreien Produkten im Lebensmittelhandel in Deutschland in den Jahren 2016 und 2017 (jeweils MAT* bis KW 12; in Millionen Euro). Chart. 16. June 2017. Statista. https://de.statista.com/statistik/daten/studie/257797/umfrage/umsatzwicklung-bei-glutenfreien-produkten-in-deutschland/. Accessed: 19 July 2019.

Lebwohl, B., Cao, Y., Zong, G., Hu, F. B., Green, P. H. R., Neugut, A. I., et al. (2017). Long term gluten consumption in adults without celiac disease and risk of coronary heart disease: Prospective cohort study. *British Medical Journal, 357,* j1892.

Lebwohl, B., Sanders, D. S., & Green, P. H. R. (2018). Coeliac disease. *Lancet, 391,* 70–81.

Lis, D. M., Fell, J. W., Ahuja, K. D. K., Kitic, C. M., & Stellingwerff, T. (2016). Commercial hype versus reality: Our current scientific understanding of gluten and athletic performance. *Current Sports Medicine Reports, 15,* 262–268.

Lobitz, R. (2018). Urgetreide – Mehr Schein als Sein? *Ernährung Im Fokus, 3–4,* 114–119.

Loponen, J., Sontag-Strohm, T., Venalainen, J., & Salovaara, H. (2007). Prolamin hydrolysis in wheat sourdoughs with differing proteolytic activities. *Journal of Agriculture and Food Chemistry, 55,* 978–984.

Losowsky, M. S. (2008). A history of coeliac disease. *Digestive Diseases, 26,* 112–120.

Louis, P., Flint, H. J., & Michel, C. (2016). How to manipulate the microbiota: Prebiotics. In A. Schwiertz (Ed.), *Microbiota of the human body* (pp. 119–142). Cham: Springer.

Marcussen, T., Sandve, S. R., Heier, L., Spannagl, M., Pfeifer, M., International Wheat Genome Sequencing Consortium, Jakobsen, K. S., Wulff, B. B., Steuernagel, B., Mayer, K. F., et al. (2014). Ancient hybridizations among the ancestral genomes of bread wheat. *Science, 345,* 1250092.

Matthes, H. (2016). Prä- und Probiotika. In A. Stallmach & M. J. G. T. Stallmach (Eds.), *Mikrobiom* (pp. 269–298). Berlin: De Gruyter.

Melini, V., & Melini, F. (2019). Gluten-free diet: Gaps and needs for a healthier diet. *Nutrients, 11,* 170.

Mintel. (2016). Anteil der glutenfreien und laktosefreien Lebensmittel an den gesamten Produktneueinführungen in Deutschland im Vergleich der Jahre 2011 und 2015. Chart. 5. August 2016. Statista. https://de.statista.com/statistik/daten/studie/587949/umfrage/anteil-glutenfreier-und-laktosefreier-lebensmittel-bei-produktneueinfuehrungen/. Accessed: 19 July 2019.

Molina-Infante, J., & Carroccio, A. (2017). Suspected nonceliac gluten sensitivity confirmed in few patients after gluten challenge in double-blind, placebo-controlled trials. *Clinical Gastroenterology and Hepatology, 15,* 339–348.

O'Connor, E. M. (2013). The role of gut microbiota in nutritional status. *Current Opinion in Clinical Nutrition and Metabolic Care, 16,* 509–516.

Ozuna, C. V., Iehisa, J. C., Gimenez, M. J., Alvarez, J. B., Sousa, C., & Barro, F. (2015). Diversification of the celiac disease alpha-gliadin complex in wheat: A 33-mer peptide with six overlapping epitopes, evolved following polyploidization. *Plant Journal, 82,* 794–805.

Plaum, P. (2019). *Chronische Obstipation als Symptom der Glutensensitivität und Zöliakie: Wie es dazu kommt, und was Patienten hilft.* Medscape Deutschland vom 12. Juni 2019. https://deutsch.medscape.com/.

Potter, M. D. E., Walker, M. M., Keely, S., & Talley, N. J. (2018). What's in a name? 'Non-coeliac gluten or wheat sensitivity': Controversies and mechanisms related to wheat and gluten causing gastrointestinal symptoms or disease. *Gut, 67,* 2073–2077.

Priyanka, P., Gayam, S., & Kupec, J. T. (2018). The role of a low fermentable oligosaccharides, disaccharides, monosaccharides, and polyol diet in nonceliac Gluten sensitivity. *Gastroenterology Research and Practice, 2018,* 1561476.

Qin, J., Li, R., Raes, J., Arumugam, M., Burgdorf, K. S., Manichanh, C., et al. (2010). A human gut microbial gene catalogue established by metagenomic sequencing. *Nature, 464,* 59–65.

Reese, I., Schäfer, C., Kleine-Tebbe, J., Ahrens, B., Bachmann, O., Ballmer-Weber, B., et al. (2018). Non-Celiac Gluten-/Wheat Sensitivity (NCGS) – A currently undefined disorder without validated diagnostic criteria and of unknown prevalence. Position statement of the task force on food allergy of the German Society of Allergology and clinical Immunology (DGAKI). *Allergo Journal International, 27,* 145–151.

Ribeiro, M., & Nunes, F. M. (2019). We might have got it wrong: Modern wheat is not more toxic for celiac patients. *Food Chemistry, 278,* 820–822.

Sapone, A., Bai, J. C., Ciacci, C., Dolinsek, J., Green, P. H. R., Hadjivassiliou, M., et al. (2012). Spectrum of gluten-related disorders: Consensus on new nomenclature and classification. *BMC Medicine, 10,* 13.

Schalk, K., Lang, C., Wieser, H., Koehler, P., & Scherf, K. A. (2017). Quantitation of the immunodominant 33-mer peptide from alpha-gliadin in wheat flours by liquid chromatography tandem mass spectrometry. *Scientific Reports, 7,* 45092.

Schalk, K., Lexhaller, B., Koehler, P., & Scherf, K. A. (2017). Isolation and characterization of gluten protein types from wheat, rye, barley and oats for use as reference materials. *PLoS ONE, 12,* e0172819.

Scherf, K. A., & Koehler, H. (2016). Wheat and gluten: Technological and health aspects. *Ernährungsumschau, 63,* 166–175.

Schuppan, D., & Zevallos, V. (2015). Wheat amylase trypsin inhibitors as nutritional activators of innate immunity. *Digestive Diseases, 33,* 260–263.

Schuppan, D., Pickert, G., Ashfaq-Khan, M., & Zevallos, V. (2015). Non-celiac wheat sensitivity: Differential diagnosis, triggers and implications. *Best Practice & Research Clinical Gastroenterology, 29,* 469–476.

Schwalb, T., Wieser, H., & Koehler, P. (2012). Studies on the gluten-specific peptidase activity of germinated grains from different cereal species and cultivars. *European Food Research and Technology, 235,* 1161–1170.

Shashikanth, N., Yeruva, S., Ong, M. L. D. M., Odenwald, M. A., Pavlyuk, R., & Turner, J. R. (2017). Epithelial organization: The gut and beyond. *Comprehensive Physiology, 7,* 1497–1518.

Shewry, P. R., Hawkesford, M. J., Piironen, V., Lampi, A. M., Gebruers, K., Boros, D., et al. (2013). Natural variation in grain composition of wheat and related cereals. *Journal of Agriculture and Food Chemistry, 61,* 8295–8303.

Sollid, L. M., Iversen, R., Steinsbo, O., Qiao, S. W., Bergseng, E., Dorum, S., et al. (2015). Small bowel, celiac disease and adaptive immunity. *Digestive Diseases, 33,* 115–121.

Statista. (2017). Warum vermeiden Sie Gluten? Chart. 30. Juni 2017. Statista. https://destat istacom/statistik/daten/studie/721875/umfrage/gruende-fuer-den-konsum-glutenfreier-nahrungsmittel-in-deutschland. Accessed: 19 July 2019.

Statista. (2019). Halten Sie sich an eine oder mehrere der folgenden Ernährungsweisen? Chart. 23. April 2019. Statista. https://destatistacom/prognosen/999784/umfrage-in-deu tschland-zu-ernaehrungsweisen. Accessed: 19 July 2019.

Tanpowpong, P., Broder-Fingert, S., Katz, A. J., & Camargo, C. A. (2012). Predictors of gluten avoidance and implementation of a gluten-free diet in children and adolescents without confirmed celiac disease. *The Journal of Pediatrics, 161,* 471–475.

Techniker Krankenkasse. (2016). Verbreitung von Nahrungsmittelunverträglichkeiten in Deutschland nach Art der Unverträglichkeit im Jahr 2016. Chart. 11. Januar 2017. Statista. https://de.statista.com/statistik/daten/studie/466795/umfrage/nahrungsmittelunvertr aeglichkeiten-nach-art-der-unvertraeglichkeit/. Accessed: 19 July 2019.

Varney, J., Barrett, J., Scarlata, K., Catsos, P., Gibson, P. R., & Muir, J. G. (2017). FOD-MAPs: Food composition, defining cutoff values and international application. *Journal of Gastroenterology and Hepatology, 32,* 53–61.

Verdu, E. F., Galipeau, H. J., & Jabri, B. (2015). Novel players in coeliac disease pathogenesis: Role of the gut microbiota. *Nature Reviews Gastroenterology & Hepatology, 12,* 497–506.

Wieser, H. *Vergleich von reinen Dinkeln und Dinkel/Weizen-Kreuzungen.* Arbeitsgemeinschaft Getreideforschung e. V. www.agfdt.de/loads/gc06/wieser.pdf. Accessed: 14 July 2019.

Wieser, H. (2007). Chemistry of gluten proteins. *Food Microbiology, 24,* 115–119.

Zeißig, S. (2016). Die physiologische Standortflora. In A. Stallmach & M. J. G. T. Vehreschild (Eds.), *Mikrobiom* (pp. 61–82). Berlin: De Gruyter.

Zevallos, V. F., Raker, V., Tenzer, S., Jimenez-Calvente, C., Ashfaq-Khan, M., Russel, N., Pickert, G., Schild, H., Steinbrink, K., & Schuppan, D. (2017). Nutritional wheat amylase-trypsin inhibitors promote intestinal inflammation via activation of myeloid cells. *Gastroenterology, 152,* 1100–1113, e1112.

Ziegler, J. U., Steiner, D., Longin, C. F. H., Wuerschum, T., Schweiggert, R. M., & Carle, R. (2016). Wheat and the irritable bowel syndrome – FODMAP levels of modern and ancient species and their retention during bread making. *Journal of Functional Foods, 25,* 257–266.

Zuidmeer, L., Goldhahn, K., Rona, R. J., Gislason, D., Madsen, C., Summers, C., et al. (2008). The prevalence of plant food allergies: A systematic review. *Journal of Allergy and Clinical Immunology, 121,* 1210–1218.

Printed in the United States
by Baker & Taylor Publisher Services